The STEAM Chasers Phases of the Moon Journal

By: Dr. Doresa A Jennings

The STEAM Chasers: Phases of the Moon Journal

THE STEAM
CHASERS

The STEAM Chasers Phases of the Moon Journal

Welcome to the STEAM Chasers Phases of the Moon Journal! Over the next thirty days, you will learn exciting facts about the moon and track the various phases of the moon from your own backyard.

You will see the moon in the following phases:

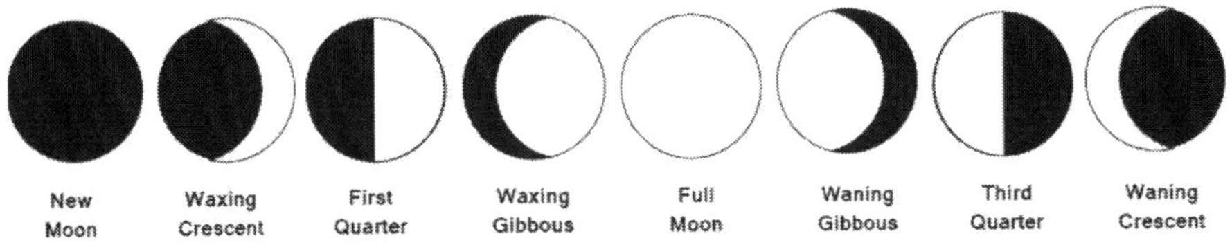

Use the grid below to practice drawing each phase listed above.

You are Now Ready to Begin!

Start by writing the date, the time of observation, and phase the moon for the night observed. Fill in your moon so that the shading matches what you observe as closely as possible. Draw on the grid any other interesting things you observe in the night sky. This can include bright stars, constellations, and planets. If anything else catches your eye, feel free to include it. These observations can be made just with your eyes, with binoculars, or with a telescope. Remember, all observations are valid. Aided (binoculars and telescopes) or unaided (using only your eyes) are both great!

Take a minute to write what you think you are going to observe over the next thirty days?

____

____

____

____

The STEAM Chasers Phases of the Moon Journal

Day 1

* On average, the Moon is 238,750 miles (384,400 km) from Earth.

Date__________________

Time__________________

Weather _______________

Moon Type ____________

Map out any constellations you saw tonight.

Note anything interesting that you saw in the sky tonight:

____

____

____

____

The STEAM Chasers Phases of the Moon Journal

Day 2

*Temperatures on the Moon can range from -200 degrees F to more than 200 degrees F.

Date________________

Time________________

Weather ______________

Moon Type ____________

Map out any constellations you saw tonight.

Note anything you saw interesting in the sky tonight.

____

____

____

____

The STEAM Chasers Phases of the Moon Journal

Day 3

*The largest known crater in our solar system is found on the Moon and is called the South Pole-Aitken.

Date________________

Time________________

Weather ______________

Moon Type ____________

Map out any constellations you saw tonight.

Note anything interesting that you saw in the sky tonight:

Day 4

*The Moon's diameter is 2,159 miles (3,475 km), roughly four times smaller than the diameter of Earth.

Date___________________

Time___________________

Weather _______________

Moon Type _____________

Map out any constellations you saw tonight.

Note anything you saw interesting in the sky tonight.

____

____

____

____

The STEAM Chasers Phases of the Moon Journal

Day 5

*So far, only 12 humans have been on the Moon - the astronauts of the Apollo missions between 1969 and 1972.

Date________________

Time________________

Weather ______________

Moon Type ____________

Map out any constellations you saw tonight.

Note anything interesting that you saw in the sky tonight:

____

____

____

____

The STEAM Chasers Phases of the Moon Journal

Day 6

*The gravity of the Moon is only 1/6 that of Earth.

Date___________________

Time___________________

Weather ________________

Moon Type _____________

Map out any constellations you saw tonight.

Note anything you saw interesting in the sky tonight.

____

____

____

____

The STEAM Chasers Phases of the Moon Journal

Day 7

*If you weighed 100 pounds on Earth, you would only weigh 16.6 pounds on the Moon.

Date________________

Time________________

Weather ______________

Moon Type ___________

Map out any constellations you saw tonight.

Note anything interesting that you saw in the sky tonight:

____

____

____

____

The STEAM Chasers Phases of the Moon Journal

Day 8

*The Moon orbits the Earth roughly every 29 days.

Date________________

Time________________

Weather _____________

Moon Type ___________

Map out any constellations you saw tonight.

Note anything you saw interesting in the sky tonight.

Day 9

*The Moon isn't actually completely round, it is shaped more like an egg!

Date________________

Time________________

Weather ______________

Moon Type ____________

Map out any constellations you saw tonight.

Note anything interesting that you saw in the sky tonight:

____

____

____

____

The STEAM Chasers Phases of the Moon Journal

Day 10

*There is no atmosphere on the Moon, so there is no wind or twilight.

Date________________

Time________________

Weather _____________

Moon Type ____________

Map out any constellations you saw tonight.

Note anything you saw interesting in the sky tonight.

____

____

____

____

Day 11

*The Moon rotates at only 10 miles (16 km) per hour; the Earth rotates at 1000 miles (1609 km) per hour.

Date_______________

Time_______________

Weather ____________

Moon Type __________

Map out any constellations you saw tonight.

Note anything interesting that you saw in the sky tonight:

The STEAM Chasers Phases of the Moon Journal

Day 12

*Tides on Earth are caused by the Moon.

Date_______________

Time_______________

Weather _____________

Moon Type ___________

Map out any constellations you saw tonight.

Note anything you saw interesting in the sky tonight.

____

____

____

____

The STEAM Chasers Phases of the Moon Journal

Day 13

*The gravitational pull of the Earth causes quakes on the Moon, called Moonquakes.

Date________________

Time________________

Weather ______________

Moon Type ___________

Map out any constellations you saw tonight.

Note anything interesting that you saw in the sky tonight:

____

____

____

____

The STEAM Chasers Phases of the Moon Journal

Day 14

*There are over 500,000 craters on the surface of the Moon.

Date____________________

Time____________________

Weather ________________

Moon Type ______________

Map out any constellations you saw tonight.

Note anything you saw interesting in the sky tonight.

The STEAM Chasers Phases of the Moon Journal

Day 15

*A lunar eclipse happens when the Earth passes between the Sun and the Moon.

Date_________________

Time_________________

Weather ______________

Moon Type ____________

Map out any constellations you saw tonight.

Note anything interesting that you saw in the sky tonight:

____

____

____

____

The STEAM Chasers Phases of the Moon Journal

Day 16

*A solar eclipse happens when Moon passes between the Sun and the Earth.

Date_________________

Time_________________

Weather ______________

Moon Type ____________

Map out any constellations you saw tonight.

Note anything you saw interesting in the sky tonight.

____

____

____

____

The STEAM Chasers Phases of the Moon Journal

Day 17

*A compass would not work on the Moon because it has no magnetic field.

Date_________________

Time_________________

Weather _______________

Moon Type ____________

Map out any constellations you saw tonight.

Note anything interesting that you saw in the sky tonight:

The STEAM Chasers Phases of the Moon Journal

Day 18

*In the far future (roughly 500 million years) the Moon and Earth will be too far away from each other for total eclipses to occur.

Date________________

Time________________

Weather ______________

Moon Type ____________

Map out any constellations you saw tonight.

Note anything you saw interesting in the sky tonight.

Day 19

*A full Moon is usually five times brighter than a half Moon.

Date__________________
Time__________________
Weather _______________
Moon Type _____________

Map out any constellations you saw tonight.

Note anything interesting that you saw in the sky tonight:

____

____

____

____

The STEAM Chasers Phases of the Moon Journal

Day 20

*Alan Shepard hit a golf ball and the Moon and so far it has never been found.

Date________________

Time________________

Weather _____________

Moon Type ___________

Map out any constellations you saw tonight.

Note anything you saw interesting in the sky tonight.

____

____

____

____

The STEAM Chasers Phases of the Moon Journal

Day 21

*Theoretically you could use your cell phone or computer on the Moon because the Moon has wifi. It is beamed from Earth.

Date________________

Time________________

Weather ______________

Moon Type ___________

Map out any constellations you saw tonight.

Note anything interesting that you saw in the sky tonight:

____

____

____

____

The STEAM Chasers Phases of the Moon Journal

Day 22

*The craters on the Moon are caused by crashes from asteroids and meteorites.

Date________________

Time________________

Weather _____________

Moon Type ___________

Map out any constellations you saw tonight.

Note anything you saw interesting in the sky tonight.

The STEAM Chasers Phases of the Moon Journal

Day 23

*The Moon is the fifth largest satellite in our solar system.

Date________________

Time________________

Weather ______________

Moon Type ___________

Map out any constellations you saw tonight.

Note anything interesting that you saw in the sky tonight:

____

____

____

____

The STEAM Chasers Phases of the Moon Journal

Day 24

*Pluto is roughly 70% the diameter of the Moon.

Date________________

Time________________

Weather ______________

Moon Type ___________

Map out any constellations you saw tonight.

Note anything you saw interesting in the sky tonight.

____

____

____

____

The STEAM Chasers Phases of the Moon Journal

Day 25

*Approximately 5285 Moons could fit inside Jupiter.

Date________________

Time________________

Weather ______________

Moon Type ____________

Map out any constellations you saw tonight.

Note anything interesting that you saw in the sky tonight:

____

____

____

____

The STEAM Chasers Phases of the Moon Journal

Day 26

*The Soviet probe Luna-2 crash-landed on the Moon in 1959.

Date_________________

Time_________________

Weather ______________

Moon Type ____________

Map out any constellations you saw tonight.

Note anything you saw interesting in the sky tonight.

The STEAM Chasers Phases of the Moon Journal

Day 27

*The Earth is over 80 times heavier than the Moon.

Date_______________

Time_______________

Weather _____________

Moon Type ___________

Map out any constellations you saw tonight.

Note anything interesting that you saw in the sky tonight:

Day 28

*There is an Outer Space Treaty which treats the Moon identical to the treatment of international waters, meaning it can only be used for peaceful purposes.

Date________________

Time________________

Weather _____________

Moon Type ___________

Map out any constellations you saw tonight.

Note anything you saw interesting in the sky tonight.

The STEAM Chasers Phases of the Moon Journal

Day 29

*The bright terrain on the Moon is called the highlands.

Date________________

Time________________

Weather _____________

Moon Type ___________

Map out any constellations you saw tonight.

Note anything interesting that you saw in the sky tonight:

____

____

____

____

The STEAM Chasers Phases of the Moon Journal

Day 30

The dark terrain on the Moon is called the lunar Maria, which is Latin for "seas".

Date________________

Time________________

Weather ______________

Moon Type ____________

Map out any constellations you saw tonight.

Note anything you saw interesting in the sky tonight.

____

____

____

____

Draw a one-month view of the phases of the moon you observed over the last thirty-days.

The STEAM Chasers Phases of the Moon Journal

Draw out the constellations you observed over the last thirty-days.

What did you learn about the moon during your month-long observation?

The STEAM Chasers Phases of the Moon Journal

Shout out your moon observation experience! What are the three things that stood out to you the most? Write your thoughts in the bubbles below.

Congratulations! You have completed your STEAM Chasers Phases of the Moon Journal! You can continue to complete this activity every month for the next eleven months and you will have an entire year of observing the moon.

In the meantime – make sure to celebrate your amazing accomplishment. You did a great job!

Sincerely,

Dr. Jennings

The STEAM Chasers Phases of the Moon Journal

Extra Observation Pages.

Date__________________
Time__________________
Weather _______________
Moon Type _____________

Map out any constellations you saw tonight.

Note anything you saw interesting in the sky tonight.

The STEAM Chasers Phases of the Moon Journal

Extra Observation Pages.

Date________________

Time________________

Weather ______________

Moon Type ____________

Map out any constellations you saw tonight.

Note anything you saw interesting in the sky tonight.

Extra Observation Pages.

Date_________________

Time_________________

Weather ______________

Moon Type ____________

Map out any constellations you saw tonight.

Note anything you saw interesting in the sky tonight.

____

____

____

____

____

The STEAM Chasers Phases of the Moon Journal

Extra Observation Pages.

Date________________

Time________________

Weather _____________

Moon Type ___________

Map out any constellations you saw tonight.

Note anything you saw interesting in the sky tonight.

The STEAM Chasers Phases of the Moon Journal

Extra Observation Pages.

Date________________

Time________________

Weather _____________

Moon Type ___________

Map out any constellations you saw tonight.

Note anything you saw interesting in the sky tonight.

____

____

____

____

____

The STEAM Chasers Phases of the Moon Journal

Extra Observation Pages.

Date________________

Time________________

Weather _____________

Moon Type ___________

Map out any constellations you saw tonight.

Note anything you saw interesting in the sky tonight.

The STEAM Chasers Phases of the Moon Journal

Extra Observation Pages.

Date_________________

Time_________________

Weather ______________

Moon Type ____________

Map out any constellations you saw tonight.

Note anything you saw interesting in the sky tonight.

The STEAM Chasers Phases of the Moon Journal

Extra Observation Pages.

Date________________

Time________________

Weather ______________

Moon Type ____________

Map out any constellations you saw tonight.

Note anything you saw interesting in the sky tonight.

The STEAM Chasers Phases of the Moon Journal

Extra Observation Pages.

Date________________

Time________________

Weather _____________

Moon Type ___________

Map out any constellations you saw tonight.

Note anything you saw interesting in the sky tonight.

Extra grid pages.

Extra grid pages.

Extra grid pages.

Extra grid page.

Extra grid pages.

Extra grid page.

Extra grid page.

Extra grid page.

Extra grid page.

Extra grid page.

Extra grid page.

Made in United States
Orlando, FL
24 November 2023

39390555R00038